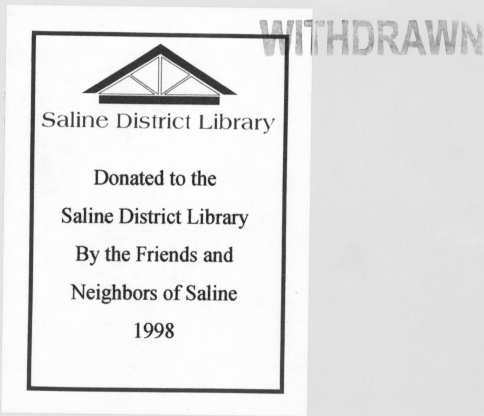

In The Wild

Elephants

Claire Robinson

Heinemann
LIBRARY

Produced by Times, Malaysia
Designed by Celia Floyd
Cover design by Lucy Smith

01 00 99 98 97
10 9 8 7 6 5 4 3 2 1

ISBN 1-57572-135-X

Library of Congress Cataloging-in-Publication Data

Robinson, Claire, 1955-
  Elephants / Claire Robinson.
    p.   cm. -- (In the wild)
  Includes bibliographical references (p.   ) and index.
  Summary: Describes the habitat, physical characteristics, and
behavior of elephants living in Africa, and more.
  ISBN 1-57572-135-X (lib. bdg.)
  1. Afrcan elephant--Juvenile literature.   [1. African elephant.
2. Elephants.]   I. Title.   II. Series: Robinson, Claire, 1955-   In
the wild.
  QL737.P98R624   1997
  599. 67' 4--dc21   97-12311
                                                         CIP
                                                          AC

Acknowledgements

The author and publishers are grateful to the following for permission to reproduce copyright
photographs:
Ardea/Joanna Van Gruisen, p.4;
Oxford Scientific Films/Martyn Colbeck, pp.5, 6, 7, 8, 9, 10, 11, 12, 13, 14, 15, 16, 17, 18, 19,
20, 21, 23;
Oxford Scientific Films/Richard Packwood, p.22.

Cover photograph: Oxford Scientific Films

Special thanks to Oxford Scientific Films

Every effort has been made to contact copyright holders of any material reproduced in this
book. Any omissions will be rectified in subsequent printings if notice is given to the publisher.

Some words are shown in bold, **like this**. You can find out what they mean by looking in
the glossary.

Contents

Elephant Relatives

There are two kinds of elephants. African elephants have large ears and long tusks.

African elephant

Asian
elephant

Asian elephants live in forests. They have high foreheads and small ears. Their tusks are usually shorter too.

What's it like to be an African elephant?

Where Elephants Live

Elephants need lots of space to travel and feed. African elephants live mainly on the hot, open **grasslands** of Africa. Some live in forests too.

This **herd** of elephants lives in Kenya.
They share the grasslands with **hoofed**
animals like zebra, giraffe, and
wildebeest.

Elephant Families

The female elephants live together. They are sisters, or mothers and daughters. They are led by the oldest female. Together they care for the babies.

Male elephants live apart from the females. These two males have traveled far to find a **mate**. They are fighting to see who will mate with her.

Trunks and Tusks

Elephants use their trunks for many things, like touching, making sounds, collecting food, and bathing. The trunk is also a nose, of course.

Tusks are an elephant's two front teeth.
Here a male is using them to strip **bark**
off a tree. Bark is good food.

Keeping Clean

Elephants need to keep their skins clean and free of **parasites**. Mud helps them to do this. They also spray themselves with water.

A dust bath helps too. This female uses
her trunk to throw a cloud of dust over
her back.

Finding Food

In the wet season, fresh green grass is easy to find and the **herd** has plenty to eat.

But in the dry season, no grass grows.
The elephants have to reach into the trees
and bushes for branches and leaves.

Eating and Drinking

This elephant gathers prickly branches with her trunk and stuffs them into her mouth. She has large, **ridged** teeth for chewing the tough food.

Water is very important to elephants.
Besides bathing in it, they need to drink
about 18 gallons of water every day.
That's enough to fill 14 buckets!

Babies

There is a new baby in the **herd**. The newborn **calf** is only one hour old. His mother helps him gently to his feet with her trunk and front foot.

He searches for his mother's milk. He throws back his tiny trunk and drinks from the **teats** between her front legs.

Growing Up

It will take about ten years for the **calf** to become an adult. Meanwhile there is plenty of time for play.

Once he is an adult, the young elephant will leave his mother's **herd** and join the adult males.

Elephant Facts

- African elephants are the biggest animals on land. They can live for up to 60 years.

- Large ears stop elephants from getting too hot. They flap their ears to cool the blood inside. Then the blood flows around their body, cooling it down.

- Elephant skin is very thick. It is lightly covered with hair.

- Elephants can hear very well. They can also make many sounds. They bellow and **trumpet**.

Glossary

bark Hard outside covering of a tree.

calf Baby elephant.

grasslands Very large areas of grass dotted with trees.

herd Large group of animals that live together, such as elephants, zebra or cattle.

hoofed Hoofed animal that has hoofs on its feet, like a horse.

mate Partner to have babies with.

mating Two animals making a baby together.

parasites Tiny animals that live on another animal's body.

ridged Not flat.

teats Parts of a female's body that give milk.

trumpet To make a loud sound, like a trumpet.

wildebeest Hoofed animal with horns.

Index

More Books To Read

Dorros, Arthur. *Elephant Families*. New York: HarperCollins, 1994.

Harrison, Virginia. *The World of Elephants*. Milwaukee: Gareth Stevens, 1989.

Petty, Kate. *Baby Animals: Elephants*. Hauppauge, NY: Barron, 1992.